DeltaScience ContentReaders™

Sound Energy

Contents

Preview the Book 2
What Is Sound? 3
    Sound Energy 4
    Sound Waves 5
    Modeling Sound Waves 6
    Sound Waves and Matter 7

How to Read Diagrams 8
Why Are Sounds Different From One Another? 9
    Frequency and Pitch 10
    Amplitude and Volume 12
    Absorption and Reflection 14

Compare and Contrast 16
How Do We Make and Hear Sound? 17
    Musical Instruments 18
    Speaking 20
    Hearing 22

Glossary 24

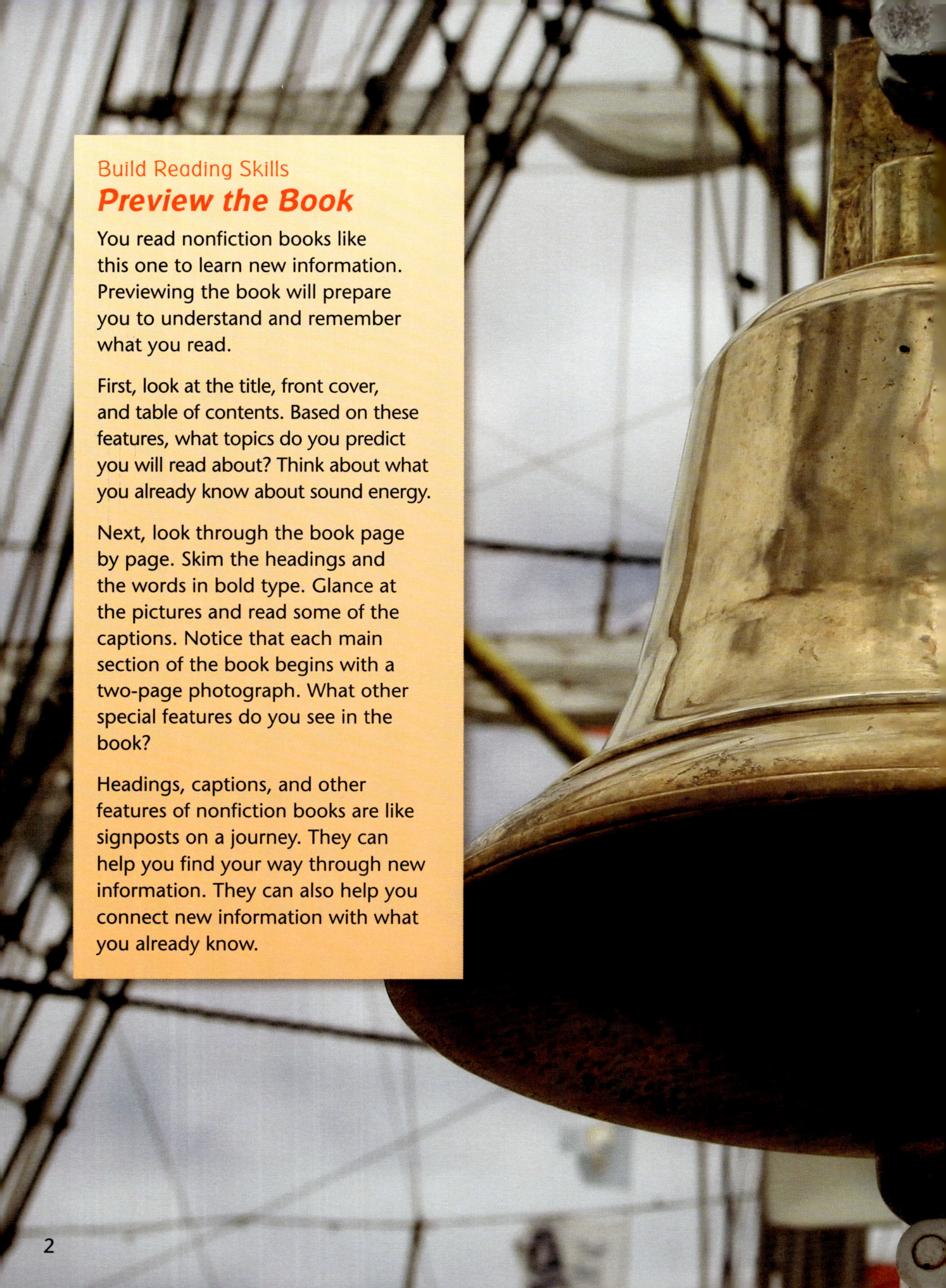

Build Reading Skills
Preview the Book

You read nonfiction books like this one to learn new information. Previewing the book will prepare you to understand and remember what you read.

First, look at the title, front cover, and table of contents. Based on these features, what topics do you predict you will read about? Think about what you already know about sound energy.

Next, look through the book page by page. Skim the headings and the words in bold type. Glance at the pictures and read some of the captions. Notice that each main section of the book begins with a two-page photograph. What other special features do you see in the book?

Headings, captions, and other features of nonfiction books are like signposts on a journey. They can help you find your way through new information. They can also help you connect new information with what you already know.

What Is Sound?

MAKE A CONNECTION
When a bell like this one rings, people nearby and far away can hear it. How do you think the bell produces sound? How do you think sound travels from an object like the bell to our ears?

FIND OUT ABOUT
- how vibrating objects produce sound waves
- how sound waves travel through matter

VOCABULARY

vibration, p. 4
compression, p. 5
sound wave, p. 5

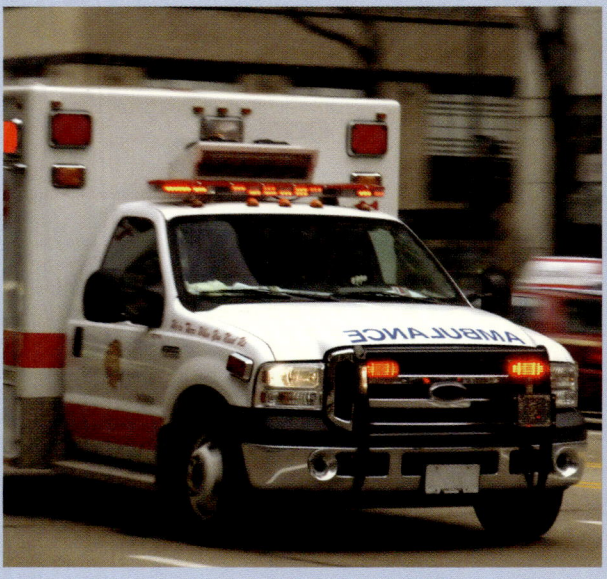

A ringing alarm clock, an ambulance siren, and a chirping bird all produce sound energy. What objects around you are producing sound energy right now?

Sound Energy

What sounds have you heard today? You might have heard a ringing alarm clock, a blaring siren, or a chirping bird. Perhaps you heard laughter, music, or conversation. All of the sounds you hear every day are the results of sound energy.

Sound energy is a form of energy we can hear. Energy is the ability to cause change, and sound energy causes a change in the motion of the particles of matter. How does the change in the motion of these particles create a sound that we can hear?

Imagine plucking a guitar string. When your finger plucks the string, the string vibrates. The string also causes other parts of the guitar to vibrate. A **vibration** is a quick back-and-forth motion. A vibrating object moves back and forth rapidly many times, causing the particles of matter around it, such as particles of air or water, to vibrate as well. These vibrations produce a sound. In fact, all sounds, including everything that you hear each day, are produced by objects vibrating.

✓ **Explain how sounds are produced.**

Strumming the guitar makes the strings vibrate. The vibrations of the guitar strings cause vibrations in the particles of air around the strings. ▶

▲ As a harp string vibrates back and forth, it creates a pattern of high-pressure areas and low-pressure areas that form sound waves.

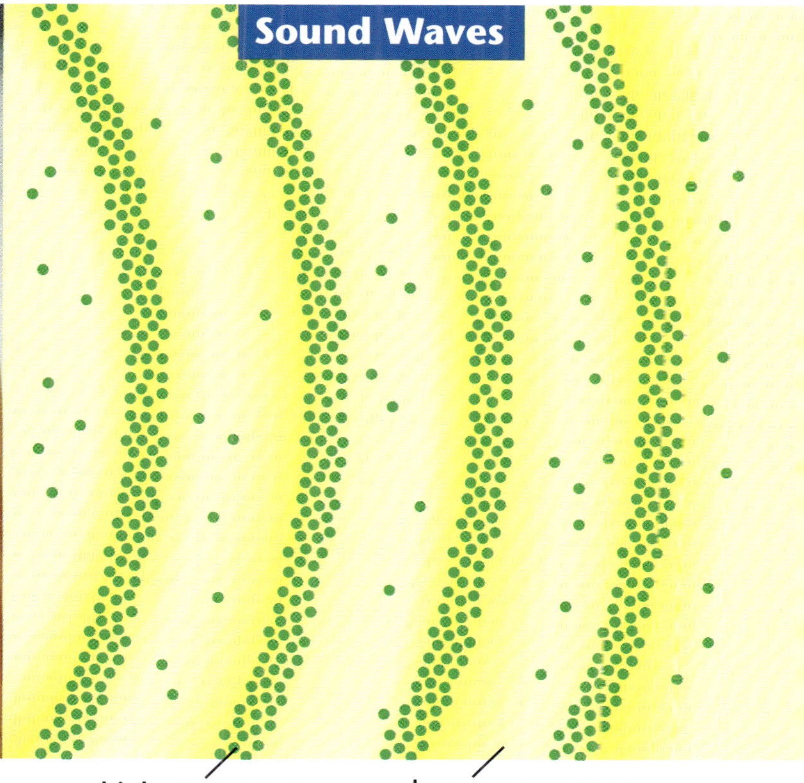

Sound Waves

high-pressure area (compression)   low-pressure area

Sound Waves

How do vibrations produce sound? Suppose you could see a vibrating guitar string or harp string in slow motion. Remember that a vibrating string moves back and forth repeatedly. Picture the string moving in each direction. As the string moves in one direction, it pushes air particles together. The bunched-up particles form a **compression**, which is an area of high pressure in which particles are squeezed together.

As the string moves in the other direction, it makes another compression. Now the space in between the two compressions has fewer air particles. It is an area of low pressure in which particles are spread out. When the string returns to the original position, it makes another compression, and so on. A vibrating string produces many areas of high and low pressure around it. The same thing occurs with other vibrating objects.

The areas of high and low pressure created by a vibrating object form repeating patterns called **sound waves**. Sound waves travel away from a vibrating object in all directions, carrying sound energy through air and other kinds of matter.

 How does hitting a drum produce sound waves?

Modeling Sound Waves

Have you ever seen waves moving across the ocean or ripples spreading over the surface of a lake? Like sound waves, water waves carry energy from one location to another. Many forms of energy travel in waves, but not all waves move in the same way.

Some waves, called *transverse waves*, cause matter to vibrate up and down as the wave travels forward. Water waves are transverse waves. Particles of water move up and down in small circles as the wave moves forward. Other waves, called *longitudinal waves*, cause matter to vibrate back and forth in the same direction the wave is traveling. Sound waves are longitudinal waves.

▲ All waves carry energy from place to place. Water waves and sound waves are alike in some ways and different in others.

You can model how sound waves move through matter using a spring toy. Stretch the spring toy along a flat surface and give one end of the spring a push. This forms a compression, an area of bunched-up coils. The compression moves along the spring from one end to the other. Sound waves carry energy away from a vibrating object in a similar way.

 How does the movement of sound waves compare to the movement of water waves?

A spring toy can be used to model the way sound waves move through matter. Energy travels along the spring in the same direction that the compressions move. ▼

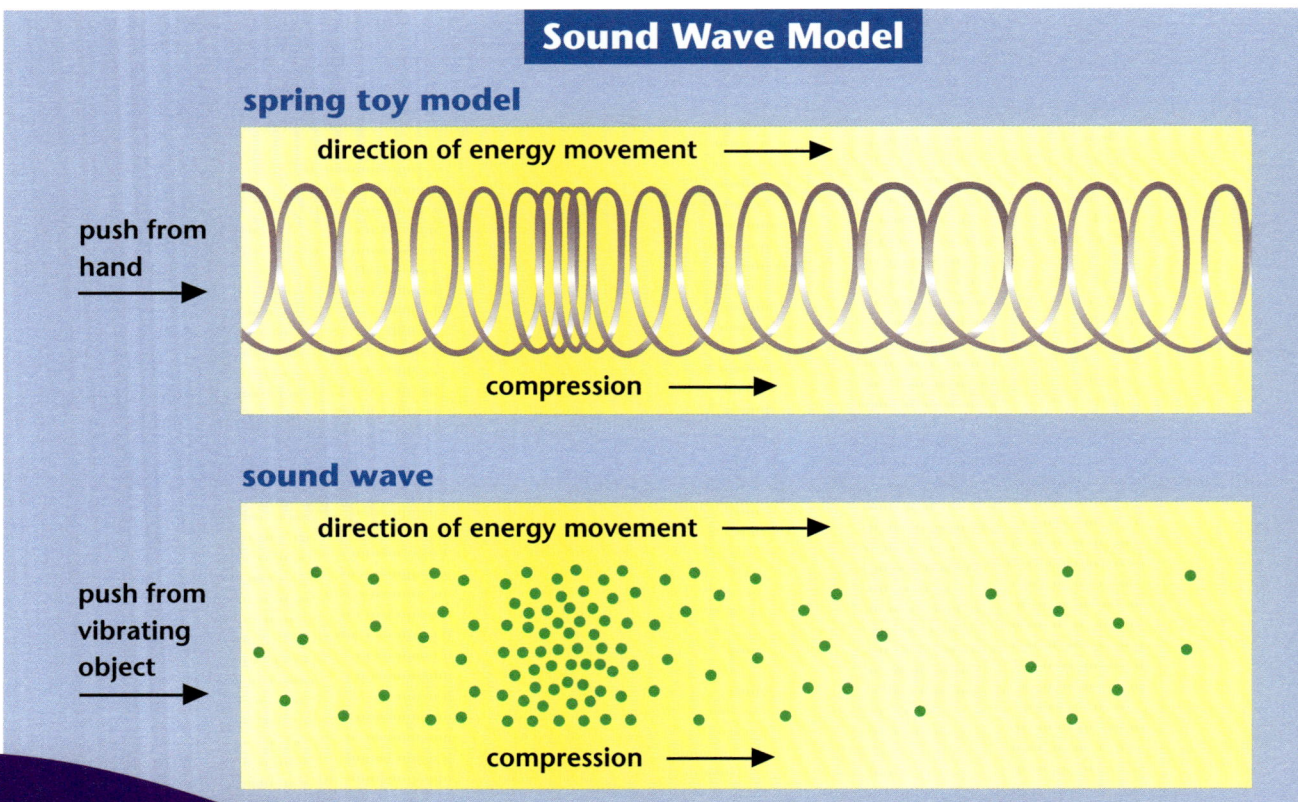

Sound Waves and Matter

Sound waves can travel through any of the three main states of matter—solid, liquid, and gas. When you talk, sound waves move through air, which is a gas. Sound waves also move through liquids, such as water, and through solids, such as walls and doors. However, sound waves cannot travel in outer space, where there is no air. Sound waves must have some kind of matter to move through.

The speed of sound depends on the kind of matter through which the sound waves are moving. Sounds travel fastest through solids, slower through liquids, and slowest of all through gases. The reason has to do with the arrangement and interaction of particles in these types of matter.

Speed of Sound

Material	Speed of Sound (meters per second)
air	346
fresh water	1,490
salt water	1,531
wood	2,000
glass	2,840
steel	5,940

▲ Sound moves at different speeds through different materials.

Temperature also affects the speed of sound. Sounds travel faster in warm air than in cold air.

 Explain why there is no sound out in space.

▲ The clicking and whistling sounds that dolphins produce travel through water, which is a liquid. Sound can also travel through gases and solids.

▲ Sound travels fast, but not as fast as light. In a thunderstorm, lightning and thunder are produced at the same time, but it takes longer for the sound to reach us.

REFLECT ON READING

Choose one heading that you noticed when you previewed this part of the book. Explain what you have learned about the topic named in the heading.

APPLY SCIENCE CONCEPTS

Choose an object that creates a sound you hear every day. Draw an illustration, using arrows and labels, that shows how the object makes sound and how the sound waves travel from the object to your ear.

Build Reading Skills
How to Read Diagrams

A **diagram** is a picture or drawing with labels. It can show how something works or how the parts of something fit together.

You will see diagrams on pages 10 and 12. As you read, think about how the diagrams help you understand the information in the text.

TIPS

Follow these guidelines when reading a diagram:

1. Read the title of the diagram and look at the picture to find out what object or event is shown.
2. Read each label and look closely at the part of the diagram it goes with.
3. Follow any arrows to understand direction or size.
4. Read the caption.
5. Explain in your own words what the diagram shows.
6. Find and reread the part of the text that discusses what the diagram shows.

A good way to understand and remember what a diagram shows is to redraw it yourself.

Why Are Sounds Different From One Another?

MAKE A CONNECTION

When a musician plays a keyboard, every key produces a different sound. What are some differences between the sounds the keys make?

FIND OUT ABOUT
- sound waves
- what makes sounds different
- what can occur when sound waves hit objects

VOCABULARY

wavelength, p. 10
frequency, p. 10
pitch, p. 11
amplitude, p. 12
volume, p. 13
absorb, p. 14
reflect, p. 14
echo, p. 15

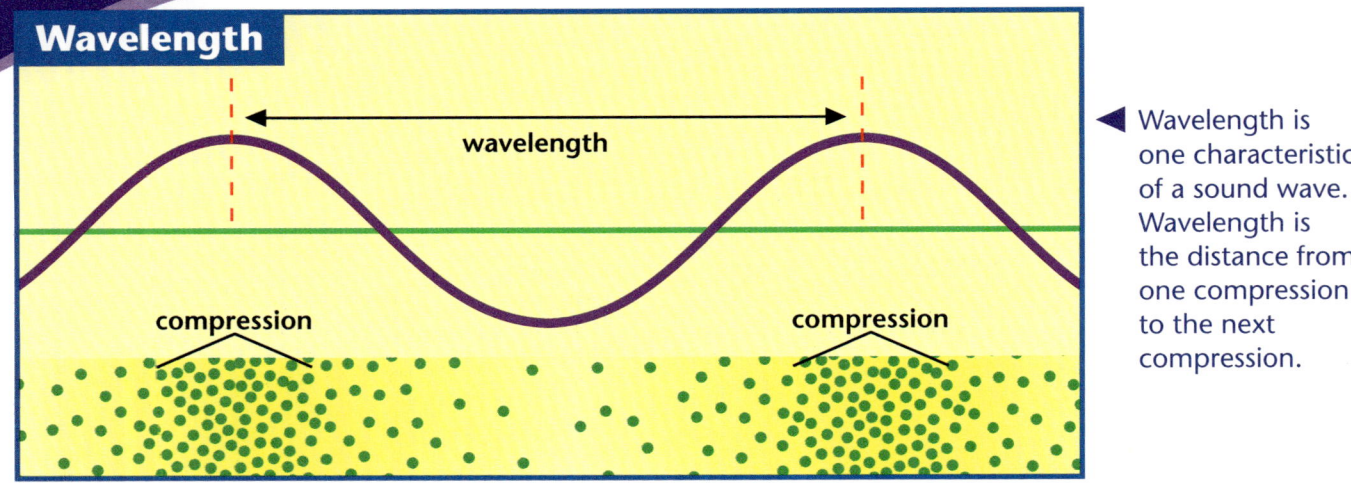

Wavelength is one characteristic of a sound wave. Wavelength is the distance from one compression to the next compression.

Frequency and Pitch

Shouting in a canyon makes a very different sound from whispering in a library. A foghorn sounds different from a flute. Why? Differences in sounds are caused by the characteristics of sound waves and how sound waves are affected by various surroundings.

The length of a wave is called its **wavelength**. Wavelength is measured from a point on one wave to the same point on the next wave. For a longitudinal wave such as a sound wave, wavelength is measured from one compression to the next. When an object vibrates quickly, the compressions are close together and the wavelength is short. When an object vibrates slowly, the compressions are far apart and the wavelength is long.

Frequency is the number of wavelengths that pass a given point each second. Each vibration produces one wavelength, so we can also think of frequency as the number of vibrations per second. The faster an object vibrates, the higher its frequency. The higher its frequency, the shorter its wavelength. Short wavelengths have a high frequency because more compressions pass a given point each second. Long wavelengths have a low frequency because only a few compressions pass a given point each second.

The longer the wavelength of a sound wave is, the lower its frequency. The shorter the wavelength, the higher the frequency. ▶

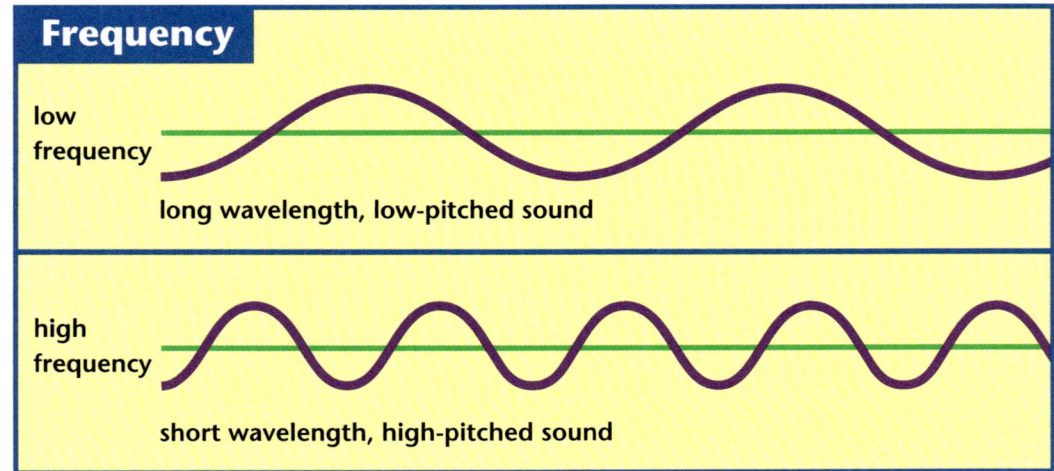

The mew of a kitten has a high pitch, while the roar of a lion has a low pitch. ▶

◀ The longer keys on a xylophone make sound waves with a longer wavelength and lower frequency. The pitch of the keys gets lower as the keys get longer.

We hear different frequencies of sound waves as differences in pitch. **Pitch** describes how high or low a sound is. A kitten's mew and the squeal of a car's brakes are high-pitched sounds. A lion's roar and the deep rumble of a tuba are low-pitched sounds.

The pitch of a sound is determined by how fast the object making the sound is vibrating. The faster an object vibrates, the higher the frequency of the sound waves it produces. High-frequency sound waves produce high-pitched sounds, and low-frequency sound waves produce low-pitched sounds.

A kitten's vocal cords vibrate quickly, producing sound waves with a short wavelength and high frequency. These high-frequency waves create sounds with high pitches. That is why a kitten makes noises that are high-pitched and squeaky.

A lion's vocal cords vibrate slowly, producing sound waves with a long wavelength and low frequency. These low-frequency waves create sounds with low pitches, so a lion's roar is a low-pitched, growling sound.

 If an object vibrates very slowly, will it make a low-pitched sound or a high-pitched sound? Explain.

▲ Loud sounds, such as the roar of a jet airliner taking off, have a great deal of energy and therefore high amplitude.

▲ A guitar can produce a loud or quiet sound, depending on how forcefully the strings are plucked.

Amplitude and Volume

Sounds can be loud or soft, depending on how much energy the sound wave has. Louder sounds have more energy than quieter sounds do. Amplitude determines the amount of energy a sound wave has.

The **amplitude** of a sound wave is how tightly pressed together the particles are in a compression. Suppose you pluck a guitar string very hard. The string will move far away from its resting position, pushing hard on the air around it and making a tight compression with high pressure. Sound waves with tighter compressions have high amplitude and therefore high energy. The higher the energy of a sound wave, the louder the sound it produces.

If you pluck a guitar string gently, it moves back and forth only a small distance. The string does not push the air as hard, so it makes looser compressions with lower pressure. Sound waves with looser compressions have low energy. Therefore, the amplitude of the sound waves is lower when you pluck the string gently than when you pluck the string hard. The lower the energy of a sound wave, the quieter the sound it produces.

Amplitude is how tightly pressed together the particles are in a compression. Sound waves with high amplitude have higher energy and higher pressure than sound waves with low amplitude do. ▶

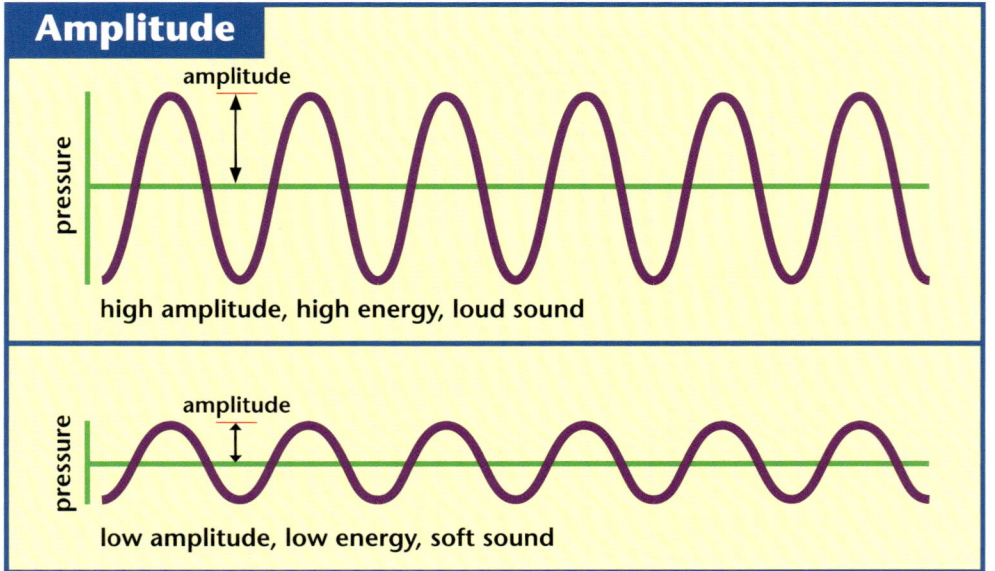

Volume is a measure of how loud or soft a sound is. The loud sounds made by jackhammers, fire engine sirens, and jet planes have a high volume. In contrast, the soft sounds made by purring cats, bubbling water fountains, and whispering people have a low volume.

Why do different sounds have different volumes? Volume is a measure of the amount of sound energy that reaches your ears, so volume is determined by the amplitude of the sound wave. High-energy, high-amplitude sound waves produce loud, or high-volume, sounds. Low-energy, low-amplitude sound waves produce soft, or low-volume, sounds.

The loud sound of a jackhammer has a high volume, while the quiet sound of a water fountain has a low volume. Volume, or loudness, is related to amplitude, or the energy of a sound wave. ▼

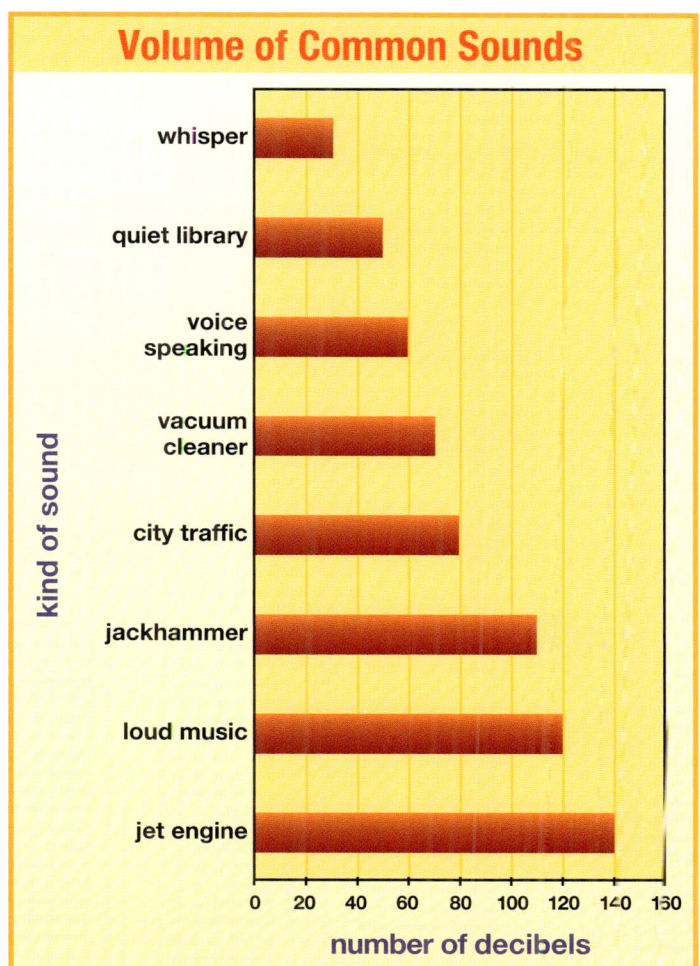

▲ The volume of sounds is measured in units called decibels.

Volume is measured in units called decibels. The greater the volume of a sound, the greater the number of decibels. For example, the loud sound of a jet engine is about 140 decibels, while the soft sound of a person whispering is about 30 decibels.

Volume is affected by the distance between the source of the sound and our ears. The volume of a sound decreases as we go farther away from the source of the sound. The volume of a sound increases as we get closer to the source of the sound.

✅ When you turn up the volume on the TV, how does the amplitude of the sound change?

Absorption and Reflection

Some materials **absorb**, or take in, most of the sound waves that strike them. Imagine walking across a floor covered with a thick rug. You might not be able to hear your footsteps, or your footsteps might seem very quiet, because soft materials, such as rugs, absorb sound waves.

In many buildings, special materials are used to make ceilings that absorb sound. This keeps sound from traveling from one room to the next. In movie theaters, thick padding on the walls is used to absorb sound. Foam is also used for sound absorption, which is why some people use foam earplugs to protect their ears while doing noisy jobs.

Other materials **reflect** most of the sound waves that strike them. When sound waves reflect, they bounce off a material. Materials that reflect sounds are usually hard and smooth, such as wood and stone. Imagine walking in hard-soled shoes on a hard wooden floor. Your footsteps would probably sound quite loud because of the reflection of sounds off the wooden floor.

▲ A rug absorbs the sounds of footsteps, but a bare wooden floor reflects these sounds.

▲ Sound-absorbing ceiling tiles are used in many kinds of buildings.

▲ Sonar is a type of sound technology that helps scientists observe and map the ocean floor.

The reflection of sound can cause an **echo**, the repetition of a sound as waves are reflected off a hard surface. Shouting into a canyon often will result in an echo. Scientists sometimes use echo technology to explore places they cannot see, such as the deep ocean.

Recall that sounds move at particular speeds depending on the material they are moving through. Scientists can release sound waves toward the bottom of the ocean. Then they can measure how long it takes for the sound waves to reflect off the ocean floor and return to them. The longer the time, the greater the distance to the bottom. This technology, called sonar, can be used to create maps of the ocean floor. Sonar stands for "**so**und **na**vigation and **r**anging."

People who design buildings consider how sounds are absorbed and reflected. For example, concert halls are made with many different materials. Some materials reflect sound toward the audience. Some materials absorb sound, keeping outside noises from interrupting a performance.

 Explain why a ball bouncing on a sidewalk sounds different from a ball bouncing on grass.

▲ In this concert hall, the padded seats and curtains absorb sound waves. The wooden floor of the stage and the angles in the ceiling reflect sound waves toward the audience.

REFLECT ON READING

Create a diagram of a sound wave with labels showing the wavelength, frequency, and amplitude. Then, with a partner, discuss the pitch and volume of the sound your sound wave would create.

APPLY SCIENCE CONCEPTS

Brainstorm some sounds that you have heard. Then create a chart to organize these sounds by pitch and volume. Label the columns *high pitch* and *low pitch*. Label the rows *high volume* and *low volume*.

Build Reading Skills
Compare and Contrast

To **compare** objects or events is to notice how they are similar. To **contrast** objects or events is to notice how they are different.

As you read this section, try to identify similarities and differences between wind instruments and percussion instruments.

TIPS

Follow these guidelines for comparing and contrasting:

- Choose two related objects or events.
- To compare them, ask, "How are they alike? What is true about both of them?"
- To contrast them, ask, "How are they different? What is true about one that is not true about the other?"

A Venn diagram can help you organize similarities and differences.

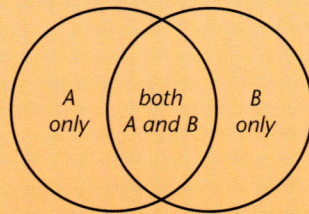

How Do We Make and Hear Sound?

MAKE A CONNECTION
Loudspeakers like these can broadcast announcements, music, or other sounds across great distances. What do you think occurs in our ears to let us hear these sounds?

FIND OUT ABOUT
- how musical instruments create sounds
- how our voices create sounds
- how we hear sounds

Musical Instruments

Musical instruments produce sounds in different ways. In fact, instruments are categorized according to how they make the vibrations that produce sound. The three types of instruments are percussion, wind, and stringed.

Pressing the valves on a clarinet changes the length of the air column inside the instrument. This changes the pitches of the notes.

maracas

cymbals

trumpet

saxophone

Percussion instruments, such as drums, xylophones, tambourines, maracas, and cymbals, vibrate and create sound waves when they are shaken or struck. The size and shape of a drum affects its pitch. Most drums have only one pitch, but the kettledrum can change pitch. A pedal tightens or loosens the kettledrum head while the drum is being played. The tighter the drum head, the higher the pitch.

A drummer can press the pedal to change the pitch of a kettledrum while it is being played. Most drums, though, have only one pitch.

Wind instruments, such as trumpets, clarinets, flutes, trombones, saxophones, and tubas, produce sound using vibrating air. A musician must blow into a wind instrument to produce sound. The column of air inside the instrument vibrates when the musician blows into it. Some wind instruments have valves or holes that a musician presses to control the length of the air column. A long column of air inside the instrument creates a low-pitched sound. A short column of air creates a high-pitched sound. The musician is able to play many different notes, each with its own pitch, by changing the length of the air column while playing.

18

banjo

harp

▲ When playing the violin or other stringed instrument, a musician can change the pitch of a sound by pressing the strings against the fingerboard.

Stringed instruments produce sounds when their strings are plucked, strummed, or moved with a bow. As the strings vibrate, they also cause the body of the instrument to vibrate and produce sound. Violins, guitars, banjos, harps, and cellos are all stringed instruments. Strings of different length, thickness, and tightness vibrate at different speeds. Long, thick, or loose strings vibrate more slowly, making low-frequency sound waves that have a low pitch. Short, thin, or tight strings vibrate more quickly, making high-frequency sound waves that have a high pitch.

Some stringed instruments, such as violins and guitars, have a fingerboard behind the strings. Musicians can play different notes by pressing a string against the fingerboard in different places. This changes the length of the part of the string that is vibrating. Changing the length of the vibrating string also changes the note the string makes. This is how musicians are able to play all the notes in a piece of music they are performing.

Musicians sometimes use tuning forks when they adjust the pitch of their instruments. When a tuning fork is tapped, it vibrates at a certain frequency, making a sound with a certain pitch. The instrument can then be tuned to match that pitch.

 Name a percussion, wind, and stringed instrument. Describe how each makes sound.

You can see the vibrations made by a tuning fork if you place it in water after it is tapped. ▶

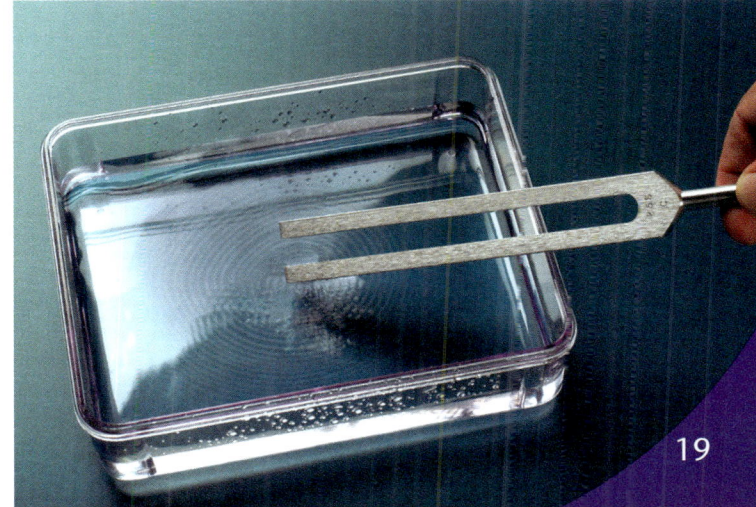

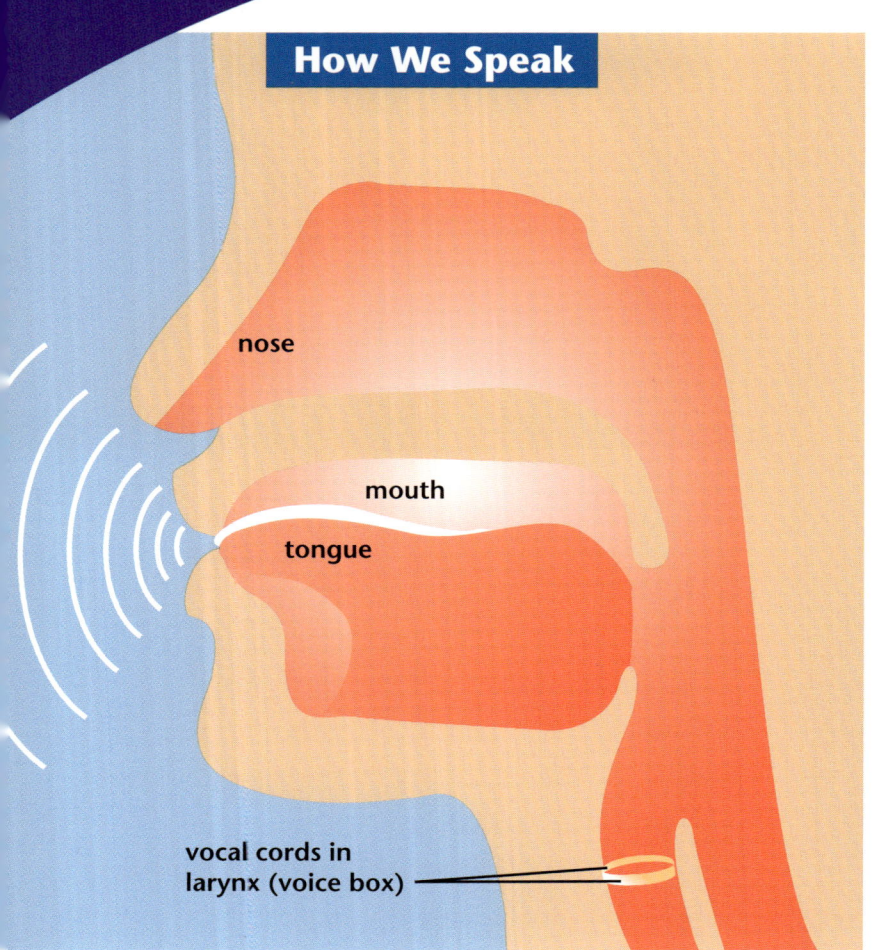

▲ The sounds of our voices are produced when air from our lungs passes our vocal cords and makes them vibrate.

▲ Vocal cords can be toned like other muscles in the body. Singers spend a lot of time doing exercises to work out their vocal cords.

Speaking

We produce sounds when we speak, shout, laugh, or sing. You know that sounds come from vibrating objects. What part of the human body vibrates to make sound? The answer is the vocal cords.

Vocal cords are folds of thin muscle tissue, similar to thin sheets of stretched rubber. Vocal cords are found in the larynx, or voice box. You can find your voice box by gently pressing your fingers to your throat while you hum or sing "Aaahhh." Move your fingers around on the front of your throat until you can feel the vibrations of your vocal cords. The diagram on this page shows the location of the voice box and vocal cords.

When we are not speaking, our vocal cords are relaxed. When we breathe out, air from the lungs makes little sound as it passes by relaxed vocal cords. When we speak or sing, our vocal cords tighten, making a small passageway for air. Air from the lungs moves between the vocal cords, causing them to vibrate and make sounds. Our teeth, tongue, and lips shape the sounds into words.

Humans can produce a wide variety of sounds. The properties of our vocal cords let us change the volume and pitch of the sounds we produce. We are able to shout or to whisper. We can hum or sing a range of high and low notes in a song.

Our vocal cords can produce sound waves with different amounts of energy. If we take a deep breath and send a rush of air over our vocal cords, they will vibrate with a lot of energy. When vocal cords vibrate with a lot of energy, loud sounds with high volumes are produced. When our vocal cords vibrate with a small amount of energy, soft sounds with low volumes are produced.

Our vocal cords can also produce sound waves with different frequencies. Tiny muscles in the larynx control the tightness of the vocal cords. The tightness determines how quickly the vocal cords vibrate. When vocal cords are tight, they vibrate quickly, making sound waves with high frequency and pitch. When vocal cords are loose, they vibrate more slowly, making sound waves with lower frequency and pitch. This is why our voices can make low-pitched and high-pitched sounds when we are speaking or singing.

 Describe how your vocal cords move and change as you sing the high and low notes of a song.

We can produce whispers, shouts, and a wide variety of other sounds with our voices. Do our vocal cords vibrate with more energy when we whisper or when we shout? ▼

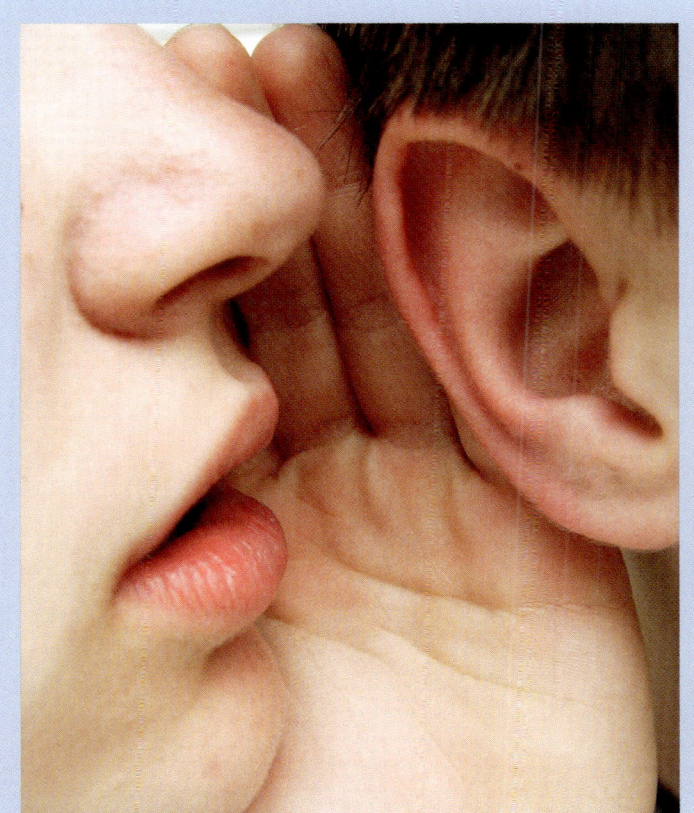

Hearing

When sound waves enter our ears, the parts of the ear work together with the brain to allow us to hear. The human ear has three sections: the outer ear, the middle ear, and the inner ear.

The outer ear is the part of the ear you can see. The funnel shape of the outer ear helps collect sound waves and directs them into the ear canal. The ear canal is a tunnel that leads to the eardrum, a thin structure located between the outer ear and the middle ear. The eardrum vibrates when sound waves hit it.

As the eardrum vibrates, it causes the hammer, anvil, and stirrup to vibrate as well. These tiny bones are found in the middle ear. The vibrations move through these bones to the inner ear, where they are changed to a form the brain can interpret.

▲ Your ears and your brain work together to allow you to hear sounds, such as your favorite music. Listening to loud music a lot can hurt your ears, so turn down the volume when you are wearing headphones.

As vibrations move from the middle ear to the inner ear, they enter the cochlea, a spiral-shaped structure that looks like a snail shell. In the cochlea, vibrations are changed to nerve signals, which then move through the auditory nerve to the brain. The brain interprets the signals as sounds.

The structures of the ear collect sounds and change them into signals that can be interpreted by the brain. ▶

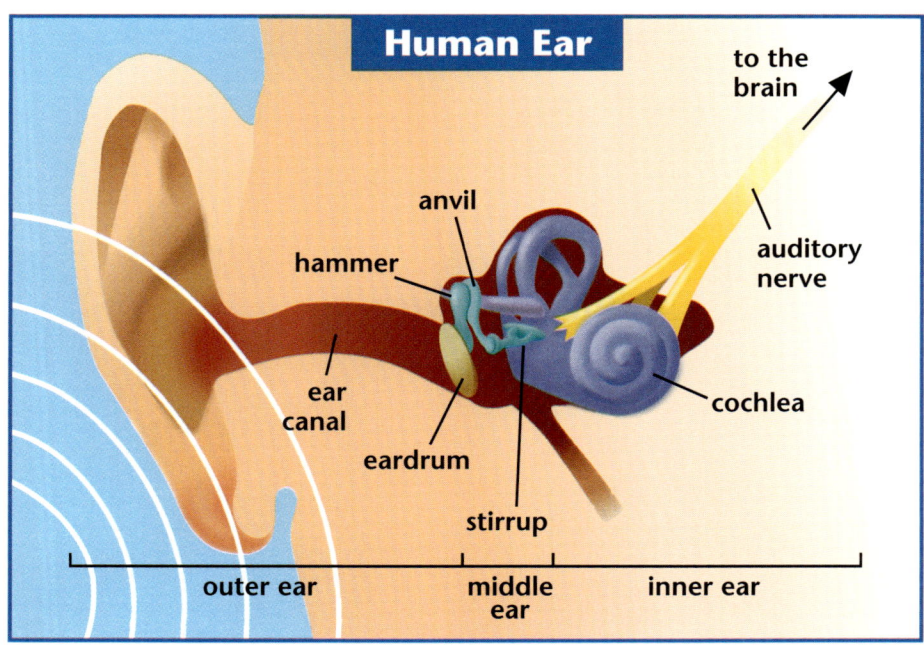

▲ Workers in noisy places must wear ear protection and take a hearing test every year. Loud sounds can damage parts of the ear.

▲ Dogs can hear sound frequencies that humans cannot hear. In fact, the high-pitched whistles used for training dogs are sometimes called silent whistles because people cannot hear them.

Damage to any part of the ear can affect hearing. Sometimes damage is caused by illness or infection. Certain parts of the ear can also be damaged by loud sounds.

Very loud sounds can damage the cochlea and the auditory nerve. This kind of damage causes hearing loss and cannot be repaired. Sounds higher than 80 decibels can cause hearing loss over a period of time, and sounds higher than 100 decibels can cause hearing loss in a very short time. For this reason, it is important to limit your exposure to very loud sounds. Using earplugs or special ear coverings can also help protect your hearing.

How does human hearing compare to that of animals? Humans can hear sound in a certain range of frequencies. Dogs and cats can hear higher-frequency sounds than humans can, and elephants can hear lower-frequency sounds than humans can.

Some animals, such as bats and dolphins, use sound to sense objects around them. Bats make a high-frequency sound that bounces off objects. The echoes help bats fly and find food in the dark. Dolphins also use *echolocation* to navigate and search for food in their ocean environment.

 What parts of the body work together to allow us to hear sounds?

REFLECT ON READING
Create a Venn diagram like the one on page 16. Use the diagram to compare and contrast how wind instruments and percussion instruments make sounds.

APPLY SCIENCE CONCEPTS
Some animals, such as rabbits and foxes, have upright ears or large ears. Write a paragraph in which you draw conclusions about the hearing of such animals.

Glossary

absorb (uhb-SORB) to take in energy **(14)**

amplitude (AM-pli-tood) a measure of how tightly pressed together the particles are in a compression **(12)**

compression (kuhm-PRESH-uhn) the part of a sound wave in which the particles are bunched close together **(5)**

echo (EK-oh) reflected sound that can be heard **(15)**

frequency (FREE-kwuhn-see) the number of wavelengths that pass through a given point each second **(10)**

pitch (PICH) how high or low a sound is **(11)**

reflect (ri-FLEKT) to strike an object and bounce off **(14)**

sound wave (SOUND WAYV) a wave that carries sound energy through matter **(5)**

vibration (vy-BRAY-shuhn) a quick back-and-forth movement **(4)**

volume (VOL-yoom) how loud or soft a sound is **(13)**

wavelength (WAYV-length) in sound waves, the distance from one compression to the next **(10)**